Binfords & Mort's hundred page, illustrated, large type FAR WESTERN CLASSICS edited by Alfred Powers

No 1—First Three Wagon Trains

No. 2—Adventures on the Columbia

ROSS COX

Adventures on the Columbia

Illustrated by

CLEVELAND ROCKWELL

and Others

BINFORDS & MORT PUBLISHERS
Portland, Oregon

Printed and bound in the United States
of America by
Binfords and Mort, Portland, Oregon

A NOTE ON ROSS COX AND HIS FAMOUS BOOK

Ross Cox, known as "the little Irishman", was only eighteen years old when he sailed in the fall of 1811 from New York for Fort Astoria to be one of John Jacob Astor's clerks in the latter's great enterprise of the Pacific Fur Company, and to stay on after the British took over with the North West Company. He had already worked a short while for Astor in New York, after arriving fresh from Dublin.

His experiences covered six years, about a year of it coming by sea and leaving overland, and five years here-and-yonder "on the banks of the Columbia and its tributary streams"; up the river nine times and down it eight; "was engaged in several encounters with the Indians; was lost fourteen days in the wilderness, and had many other extraordinary escapes".

His lost adventure came about from his lagging so much at the end of the almost mile-long fur train that the short-tempered leader rode back to prod him up and, finding the laggard sassy, to strike him with his whip. Whereupon Cox stayed behind altogether, and had such a time of it for two weeks alone in the wilderness as to make the most notable lost story of the Pacific Northwest.

He waited fourteen years before publishing the narrative of his five years on the Columbia. The reduction in the present volume to a hundred pages is from 760 pages in the two volumes of his work issued in 1831, so popular as to reach a third edition the next year. The original printing is scarce and very expensive; this condensation makes easily available the substantial portions of its excitement and of its vivid pictures of the great stream's drainage area a century and a half ago.

Cox did not become a squaw man, at least not by his own confession. In 1919, a year after he returned home to Dublin, *he married Hannah Cummings and by her had a large family. For a number of years he was Irish correspondent for a London newspaper. He was also still a clerk but he seemed always to manage to clerk where things happened—it was now in the* Dublin *Police Office. He died in 1853 at the age of sixty.*

Adventures on the Columbia *is the second volume in Binfords & Mort's series of large-type condensations of previously hard-to-get Pacific Coast classics. The illustrations are by Cleveland Rockwell, a noted Oregon artist, with additional pictures by other well-known artists, showing the river at a later period but unchanged in the views selected from the time Ross Cox saw it from 1812 to 1817.*

1

VOYAGE TO THE COLUMBIA

On Thursday the 17th of October, 1811, we sailed from New York. On the 25th of March, at daybreak, we made the island of Owhyee, the largest in the group of the Sandwich Islands (Hawaiian Islands). It was the captain's original intention to stop at this place for supplies; but we were informed by some natives who came off in canoes that Tamaahmaah, the king, then resided in Whoahoo.

On the 26th at noon we came to anchor outside of the bar in Whytetee bay. After he had arrived on the deck, Tamaahmaah shook hands in the most condescending manner with everyone he met between the cabin and the gangway, exclaiming to each person, "*Aroah, Aroah nuee* . . . I love you, I love you much."

As we intended to engage some of the natives for the Company's service at the Columbia, and as the captain also required some to assist in working the ship (several of the crew being indifferent sailors), he demanded permission from Tamaahmaah to engage the number that should be deemed necessary: this was at once granted. On the intelligence being announced, the vessel was crowded with numbers, all offering to "take on". We selected twenty-six of the most able-bodied. We agreed to pay each man ten dollars a month, and a suit of clothes annually. An old experienced islander, who was called Boatswain Tom, was engaged to command them. He got fifteen dollars a month, and was to have sole control of his countrymen.

On the 5th of April we got all our supplies on board. They consisted of sixty hogs, two boats full of sugar-cane to feed them, some thousand cocoanuts, with as much bananas, plantains, taro, melons, etc., as could be conveniently stowed in the ship. The following morning, Tuesday, April the 6th, we weighed anchor and set sail for the Columbia.

The addition we received in livestock, joined to the cargo of fruit, etc., lumbered our deck greatly, and annoyed the crew in working the ship. When any number of the natives were wanted to perform a particular duty, word was passed to Bos'n Tom. The moment he gave "the dreadful word" it was followed by a horrid yell. With a rope's end he laid on the back and shoulders of every poor devil who did not happen to be as alert as he wished, accompanied by a laughable *melange* of curses in broken English and imprecations in his own language.

On the 5th of May (1812) we had the happiness of beholding the entrance of the long-wished-for Columbia. Light, baffling winds, joined to the captain's timidity, obliged us to stand off and on until the 8th, on which day we descried a white flag hoisted on Cape Disappointment, the northern extremity of the land at the entrance of the river. A large fire was also kept burning on the cape at night, which served as a beacon.

A dangerous bar runs across the mouth of the Columbia. The channel for crossing it is on the northern side close to the cape, and is very narrow. Thence to the opposite point on the southern side, which is called Point Adams, extends a chain or reef of rocks and sand-banks, over which the dreadful roaring of the mighty waters of the Columbia, in forcing their passage to the ocean, is heard for miles distant.

Early on the morning of the 9th Mr. Rhodes was ordered out in the cutter, on the perilous duty of sounding the channel of the bar, and placing the buoys necessary for the safe guidance of the ship.

While he was performing this duty we fired several guns; and, about ten o'clock in the morning, we were delighted with hearing the report of three cannon from the shore in answer to ours. Towards noon an Indian canoe was discovered making for us, and a few moments after a barge was perceived following it. Various were the hopes and fears by which we were agitated, as we waited in anxious expectation the arrival of the strangers from whom we were to learn the fate of our predecessors, and of the party who had crossed the continent.

Vague rumors had reached the Sandwich Islands from a costing vessel that the *Tonquin* had been cut off by the Indians and every soul on board destroyed; and, since we came in sight of the river, the captain's ominous forebodings had almost prepared the weaker part of our people to hear that some dreadful fatality had befallen our infant establishment.

Not even the sound of the cannon and the sight of the flag and the fire on the cape were proofs strong enough to shake his doubts. He was too well acquainted with Indian cunning and treachery to be deceived by such appearances. It was possible enough that the savages might have surprised the fort, murdered its inmates, seized the property, fired the cannon to induce us to cross the bar, which, when once effected, they could easily cut us off before we could get out again. He even carried his caution so far as to order a party of armed men to be in readiness to receive our visitors.

The canoe arrived first alongside. In it was an old Indian, with six others, nearly naked, and the most repulsive looking beings that ever disgraced the fair form of humanity. The only intelligence we could obtain from them was that the people in the barge were white like ourselves, and had a house on shore.

A few minutes afterwards it came alongside, and dissipated all our fearful dreams of murder, and we had the delightful, the inexpressible pleasure of shaking hands with Messrs. Duncan M'Dougall and Donald M'Lennan; the former a partner and the latter a clerk of the Company, with eight Canadian boatmen.

They informed us that on receiving intelligence the day before from the Indians that a ship was off the river, they came down from the fort, a distance of twelve miles, to Cape Disappointment, on which they hoisted the flag we had seen, and set fire to several trees to serve in lieu of a lighthouse.

The tide was now making in, and as Mr.

Rhodes had returned from placing the buoys, Mr. M'Lennan, who was well acquainted with the channel, took charge of the ship as pilot. At half past two P.M. we crossed the bar, on which we stuck twice without sustaining any injury. Shortly after, we dropped anchor in Baker's Bay, after a tedious voyage of six months and twenty-six days.

Mr. M'Dougall informed us that the one-eyed Indian who had preceded him in the canoe was the principal chief of the Chinook nation, who reside on the northern side of the river near its mouth; that his name was Comcomly and that he was much attached to the whites. We therefore made him a present, and gave some trifling articles to his attendants, after which they departed.

2

FORT ASTORIA IN 1812

AFTER THE VESSEL WAS SECURELY MOORED we took our leave of the good ship *Beaverin* on which we had traveled upwards of twenty thousand miles.

In the evening we arrived at the Company's establishment, which was called Fort Astoria in honor of Mr. Astor. Here we found five proprietors, nine clerks, and ninety artisans and canoe-men or, as they are commonly called in the Indian country, *voyageurs*. We brought an addition of thirty-six, including the islanders; so that our muster roll amounted to one hundred and forty men.

[Here Cox takes thirty-one pages to tell of the ill-fated *Tonquin* and the hardships of the Astor overlanders under Wilson Price Hunt.]

The spot selected for the fort was on a handsome eminence called *Point George*, which commanded an extensive view of the majestic Columbia in front, bounded by the bold and thickly-wooded northern shore.

On the right, about three miles distant, a long, high and rocky peninsula covered with timber, called *Tongue Point,* extended a considerable distance into the river from the southern side with which it was connected by a narrow neck of land.

On the extreme left *Cape Disappointment,* with the bar and its terrific chain of breakers, were distinctly visible.

The buildings consisted of apartments for the proprietors and clerks, with a capacious dining-hall for both, extensive warehouses for the trading goods and furs, a provision store, a trading shop, smith's forge, carpenters workshop, etc., the whole surrounded by stockades forming a square and reaching about fifteen feet over the ground. A gallery ran around the stockades, in which loopholes were pierced sufficiently large for musketry. Two strong bastions built of logs commanded the four sides of the square. Each bastion had two stories in which a number of chosen men slept every night. A six-pounder was placed in the lower story, and they were both well provided with small arms.

Immediately in front of the fort was a gentle declivity sloping down to the river's side, which had been turned into an excellent kitchen garden. A few hundred yards to the left, a tolerable wharf had been run out, by which *bateaux* and boats were enabled at low water to land their cargoes without sustaining any damage.

An impenetrable forest of gigantic pine rose in the rear; and the ground was covered with a thick underwood of briar and huckleberry, intermingled with fern and honeysuckle.

Numbers of the natives crowded in and about the fort. They were most uncouth-looking objects, not strongly calculated to impress us with a favorable opinion of aboriginal beauty or the purity of Indian manners.

A few of the men were partially covered, but the greater number were unannoyed by vestments of any description. Their eyes were black, piercing, and treacherous; their ears slit up and ornamented with strings of beads; the cartilage of their nostrils perforated and adorned with pieces of *hyaquau* placed horizontally. Their heads presented an inclined plane from the crown to the upper part of the nose, totally unlike our European rotundity of cranium; and their bodies besmeared with whale oil, gave them an appearance horribly disgusting.

Then the women—O ye gods! With the same auricular, olfactory, and craniological peculiarities, they exhibited loose hanging breasts, short dirty teeth, skin saturated with blubber, bandy legs, and a waddling gait. Their only dress consisted of a kind of petticoat, or rather kilt, formed of small strands of cedar bark twisted into cords and reaching from the waist to the knee. This covering in calm weather, or in erect position, served all the purposes of concealment; but in a breeze formed a miserable shield in defense of decency. Worse than all, their repulsive familiarities rendered them objects unsupportably odious.

From these ugly specimens of mortality we turned with pleasure to contemplate the productions of their country, amongst the most wonderful of which are the fir trees. The largest species grow to an immense size. One immediately behind the fort at the height of ten feet from the surface of the earth measured forty-six feet in circumference! The trunk of this tree had about one hundred and fifty feet free from branches. Its top had been some time before blasted by lightning. To judge by comparison, its height when perfect must have exceeded three hundred feet!

Our table was daily supplied with elk, wild

fowl, and fish. Of the last we feasted on the royal sturgeon, which is here large, white, and firm; unrivalled salmon; and abundance of the sweet little anchovy, which is taken in such quantities by the Indians that we have seen their houses garnished with several hundred strings of them, dry and drying. We had them generally twice a day, and in a few weeks we got such a surfeit that few of us for years afterwards tasted an anchovy (smelt).

We remained upwards of six weeks at the fort, preparing for our grand expedition into the interior. During this period I went on several short excursions to the villages of various tribes up the river and about the bay.

We also visited Fort Clatsop, the place where Captains Lewis and Clark spent the winter of 1805-6. The logs of the house were still standing, and marked with the names of several of their party.

3

RATTLESNAKES

On the 29th of June, 1812, we took our departure from Astoria for the interior.

Our party consisted of three proprietors, nine clerks, fifty-five Canadians, twenty Sandwich islanders, and Messrs. Crooks, M'Lennan, and R. Stuart. We traveled in *bateaux* and light-built wooden canoes.

Our lading consisted of guns and ammunition, spears, hatchets, knives, beaver traps, copper and brass kettles, white and green blankets, blue, green, and red cloths, calicoes, beads, rings, thimbles, hawk-bells; and our provisions of beef, pork, flour, rice, biscuits, tea, sugar, with a moderate quantity of rum, and wine. The soft and hard goods were secured in bales and boxes, and the liquids in kegs holding on an average nine gallons. From thirty to forty of these packages and kegs were placed in each vessel.

The Columbia is a noble river, uninterrupted by rapids for one hundred and seventy miles. It is seldom less than a mile wide; but in some places its breadth varies from two to five miles. The river, up to the rapids, is covered with several islands from one to three miles in length.

We arrived on the evening of the 4th [July] at the foot of the first rapids, where we encamped. It was arranged that five officers should remain at each end of the portage, and the remainder, with twenty-five men, be stationed at short distances from each other. Its length was between three and four miles, and the path was narrow and dangerous. We only made one half of the portage the first day, and encamped near an old village.

In the course of the day, in the most gloomy part of the wood, we passed a cemetery. There were nine shallow excavations. They contained numbers of dead bodies, all carefully enveloped in mats and skins. Several poles were attached to these burial places, on which were suspended robes, pieces of cloth, kettles, bags of trinkets, baskets of roots, wooden bowls, and several ornaments; all of which the survivors believed their departed friends would require in the next world. Their veneration is so great for these offerings that it is deemed sacrilege to pilfer one of them; and, although these Indians are not remarkable for scrupulous honesty, I believe no temptation would induce them to touch these articles. Several of the boards are carved and painted with rude representations of men, bears, wolves, and animals unknown. Some in green, others in white and red, and all most hideously unlike nature.

We commenced proceedings at four o'clock on the morning of the 6th and finished the portage about two in the afternoon. On the evening of the 8th we reached the foot of the narrows, or, as the Canadians call them *les dalles*.

The river from the first rapids to the narrows is broad, deep, and rapid, with several sunken rocks scattered here and there, which often injure the canoes. The Columbia, at the narrows, for upwards of three miles is compressed into a narrow channel not exceeding sixty or seventy yards wide; the whole of which is a succession of boiling whirlpools. Above this channel for four or five miles the river is one deep rapid, at the upper end of which a large mass of high black rock stretches across from the north side and nearly joins a similar mass on the south. They are divided by a strait not exceeding fifty yards wide. Through this narrow channel, for upwards of half a mile, the immense waters of the Columbia are one mass of foam, and force their headlong course with a frightful impetuosity, which cannot at any time be contemplated without producing a painful giddiness.

We were obliged to carry all our lading from the lower to the upper narrows, nearly nine miles. The canoes were dragged up part of the space between the narrows. The laborious undertaking occupied two entire days.

About four or five miles above the fall, a high rocky island three miles in length lies in the center of the river, on which the Indians were employed drying salmon, great quantities of which were cured and piled under broad boards in stacks. We encamped on the north side opposite the island, and were visited by some Indians, from whom we purchased salmon.

The day after we stopped about one o'clock at a village, where we purchased five horses intended for the kettle.

A curious incident occurred at this spot to one of our men named La Course, which was nearly proving fatal.

This man had stretched himself on the ground, after the fatigue of the day, with his head resting on a small package of goods, and quickly fell asleep. While in this situation I passed him, and was almost petrified at seeing a large rattlesnake moving from his side to his left breast.

My first impulse was to alarm La Course; but an old Canadian whom I beckoned to the spot requested me to make no noise, alleging it would merely cross the body and go away. He was mistaken; for, on reaching the man's left shoulder, the serpent deliberately coiled himself, but did not appear to meditate an attack.

Having made signs to several others, who joined us, it was determined that two men should advance a little in front, to divert the attention of the snake, while one should approach La Course behind, and with a long stick endeavor to remove it from his body.

The snake, on observing the men advance in front, instantly raised its head, darted out its forked tongue, and shook its rattles; all indications of anger. Everyone was now in a state of feverish agitation as to the fate of poor La Course, who still lay slumbering, unconscious of his danger. The man behind, who had procured a stick seven feet in length, suddenly placed one end of it under the coiled reptile, and succeeded in pitching it upwards of ten feet from the man's body.

A shout of joy was the first intimation La

Course received of his wonderful escape. In the meantime the man with the stick pursued the snake, which he killed.

It was three feet six inches long; and eleven years old, which I need not inform my readers was easily ascertained by the number of rattles.

A general search was then commenced about the encampment. Under several rocks we found upwards of fifty of them, all of which we destroyed. There is no danger attending their destruction, provided a person has a long pliant stick, and does not approach them nearer than their length, for they cannot spring beyond it, and seldom act on the offensive except when closely pursued. They have a strong repugnance to the smell of tobacco, in consequence of which we opened a bale of it, and strewed a quantity of loose leaves about the tents, by which means we avoided their visits during the night.

[*This was Cox's first but by no means his last acquaintance with these ill-famed reptiles. On his lone trip described in the next chapter they were a constant addition to his dismays, and when he was stationed four years later at Fort Okanogan he found that neighborhood infested with them*].

The point of land upon which the fort is built is formed by the junction of the Oakinagin River with the Columbia. At the upper end is a chain of hills.

Rattlesnakes abound beyond these hills and on the opposite sides of the Oakinagan and Columbia Rivers. They are also found on both sides of the Columbia, below its junction with the former stream, but it is a curious fact that on the point itself, that is, from the rocks to the confluence of the two rivers, a rattlesnake has never yet been seen. The Indians are unable to account for this peculiarity. As we never read of St. Patrick having visited that part of the world, we were equally at a loss to divine the cause. The soil is dry and rather sandy, and does not materially differ from that of the surrounding country.

The rattlesnakes were very numerous about the place where the men were cutting timber. I have seen some of our Canadians eat them repeatedly! The flesh is very white and they assured me had a delicious taste.

Their manner of dressing them is simple. They at first skin the snake in the same manner as we do eels, after which they run through the body a small stick, one end of which is planted in the ground leaning towards the fire. By turning this *brochet* occasionally the snake is shortly roasted.

Great caution, however, is required in killing a snake for eating. If the first blow fails, or only partially stuns him, he instantly bites himself in different parts of the body, which thereby becomes poisoned and would prove fatal to any person who should partake of it. The best method is to wait until he begins to uncoil and stretches out the body preparatory to a spring, when, if a steady aim be taken with a stick about six feet long, it seldom fails to kill with the first blow.

4

LOST

ABOVE LEWIS RIVER [the Snake] the Columbia runs in a northerly direction: below it in a westerly. We remained here three days purchasing horses for our journey inland.

Mr. David Stuart and party proceeded in their canoes up the Columbia to the trading establishment which he had formed at Oakinagan River, which falls into the Columbia about two hundred and eighty miles above this place.

Mr. Donald M'Kenzie and his party proceeded up Lewis River in order to establish a trading post on the upper parts of it, or in the country of the Snake Indians.

Having purchased twenty-five horses, we took our departure on the 3rd of August, and proceeded up Lewis River; some on land with the horses, but the greater part still in the canoes. The water was very high and rapid, and in many places the banks steep and shelving, which made the process of dragging up the canoes very difficult. Poling was quite impossible; for on the off or outer side the men could not find bottom with their poles.

On the 15th of August, at five A.M. we took our departure from Lewis River. Our destination was fixed for the Spokan tribe of Indians, about one hundred and fifty miles from Lewis River in a northeast direction.

On the 17th of August we left our encampment a little after four A. M. During the forenoon the sun was intensely hot. We got no water, however, until twelve o'clock, when we arrived in a small valley of the most delightful verdure, through which ran a clear stream over a pebbly bottom. The horses were immediately turned loose and orders were given not to catch them until two o'clock.

After walking and riding eight hours, I need not say we made a hearty breakfast; after which I wandered some distance along the banks of the rivulet in search of cherries, and came to a sweet little arbor formed by sumach and cherry trees. It was a charming spot. I fell into a kind of pleasing, soothing reverie, which, joined to the morning's fatigue, gradually sealed my eyelids. Imagine my feelings when I awoke in the evening, I think it was about five o'clock from the declining appearance of the sun. All was calm and silent as the grave. I ran to the place where the men had made their fire. All, all were gone. Not a vestige of man or horse appeared in the valley.

My senses almost failed me. I called out in vain in every direction until I became hoarse. I could no longer conceal from myself the dreadful truth that I was alone in a wild, uninhabited country. without horse or arms, and destitute of covering.

To ascertain the direction which the party had taken, I set about examining the ground. At the northeast point of the valley I discovered the tracks of horses' feet, which I followed for some time and which led to a chain of small hills with a rocky, gravelly bottom on which the hoofs made no impression. Having thus lost the tracks, I ascertained the highest of the hills, but saw no sign of the party.

The evening was now closing fast, and a heavy dew commenced falling. The whole of my clothes consisted merely of a gingham shirt, nankeen trousers, and a pair of light leather moccasins, much worn. In consequence of the heat, I had taken off my coat and placed it on one of the loaded horses; and one of the men had charge of my fowling-piece. I was even without my hat, for in the agitated state of my mind on awakening I had left it behind, and had advanced too far to think of returning for it.

At some distance on my left I observed a field of high, strong grass, to which I proceeded. After pulling enough to place under and over me, I recommended myself to the Almighty and fell asleep.

On the 18th I arose with the sun, quite wet and chilly, the heavy dew having completely saturated my flimsy covering.

Late in the evening I observed, about a mile distant, two horsemen galloping in an easterly direction. From their dress I knew they belonged to our party. I instantly ran to a hillock, and called out in a voice to which hunger had imparted a supernatural shrillness; but they galloped on. I then shook off my shirt, which I waved in a conspicuous manner over my head, accompanied by the most frantic cries; still they continued on.

I ran towards the direction they were galloping, despair adding wings to my flight. Rocks, stubble, and brushwood were passed with the speed of a hunted antelope—but to no purpose; for on arriving at the place where I imagined a pathway would have brought me into their track, I was completely at fault.

It was now nearly dark. I had eaten nothing since noon of the preceding day. Faint with hunger and fatigue, I threw myself on the grass. When I heard a small rustling noise behind me, I turned around and, with horror, beheld a large rattlesnake cooling himself in the evening shade. I instantly retreated, on observing which he coiled himself. Having obtained a large stone, I advanced slowly on him and, taking a proper aim, dashed it with all my force on the reptile's head, which I buried in the ground beneath the stone.

The late race had completely worn out the thin soles of my moccasins. My feet in consequence became much swollen. As night advanced I was obliged to look out for a place to sleep. My exertions in pulling the long, coarse grass nearly rendered my hands useless by severely cutting all the joints of the fingers.

I rose before the sun. I at first felt very hungry, but after walking a few miles and taking a drink of water I got a little refreshed. The scorching influence of the sun obliged me to stop for some hours in the day; during which I made several ineffectual attempts to construct a covering for my head. At times I thought my brain was on fire from the dreadful effects of the heat.

I got no fruit those two days, and towards evening felt very weak from the want of nourishment, having been forty-eight hours without food. To make my situation more annoying, I slept that evening on the banks of a pretty lake, the inhabitants of which would have done honor to a royal table. With what an evil eye and a murderous heart did I regard the stately goose and the plump waddling duck as they sported on the water. Even with a pocket pistol I could have done execution among them.

The state of my fingers prevented me from obtaining the covering of grass which I had the two preceding nights. On this evening I had no shelter whatever to protect me from the heavy dew.

On the following day, the 20th, my course lay through a country more diversified by wood and water. The rattlesnakes were very numerous this day, with horned lizards and grasshoppers. The latter kept me in a constant state of feverish alarm from the similarity of the noise made by their wings to the sound of the rattles of the snake when preparing to dart on its prey.

I suffered severely during the day from hunger, and was obliged to chew grass occasionally, which allayed it a little.

Late in the evening I arrived at a lake upwards of two miles long. There was an abundant supply of wild cherries on which I made a hearty supper. During the night the howling of wolves and growling of bears broke in terribly on my slumbers. Next morning I observed on the opposite bank the entrance of a large and apparently deep cavern, from which I judged some of the preceding night's music had issued.

I now determined to make short journeys for two or three days in different directions in the hope of falling on some fresh horse-tracks, to return each night to the lake where I was at least certain of procuring cherries and water. I set out early in a southerly direction through a wild, barren country. I had armed myself with a long stick, with which during the day I killed several rattlesnakes.

Having discovered no fresh tracks, I returned late in the evening hungry and thirsty. I collected a heap of stones from the water side; and just as I was lying down observed a wolf emerge from the opposite cavern. I threw some stones at him, one of which struck him on the leg. He retired yelling into his den. After waiting some time in fearful suspense to see if he would reappear, I threw myself on the ground and fell asleep; but, like the night before, it was broken by the same unsocial noise and for upwards of two hours I sat up

(Left) The way Cleveland Rockwell saw Multnomah Falls in 1882, practically unchanged then from Cox's view 60 years before.

(Right) Beacon Rock, formerly known as Castle Rock, one of the earth's biggest monoliths, its precincts all civilized now even to Coca Cola.

Chinook canoe and squaw. During his first six weeks at Fort Astoria Cox frequently visited the Indians, whose "heads presented an inclined plane from the crown to the upper part of the nose."

Mt. Hood, a changeless eyeful of white and height to those ascending the Columbia. Cox, on his first trip in June, 1812, having plenty of paddlers for his bateau, could just sit and look.

waiting in anxious expectation the return of daylight.

My excursion to the southward having proved abortive, I now resolved to try the east. I had to penetrate a country full of "dark woods and rankling wilds". My feet too were uncovered and, from the thorns of the various prickly plants, were much lacerated. In consequence, on returning to my late bivouack, I was obliged to shorten the legs of my trousers to procure bandages for them. The wolf did not make his appearance; but that night I got occasional starts from several of his brethren.

On the morning of the 23rd, having been unsuccessful the two preceding days, I determined to shape my course due north, and if possible not return to the lake. During the day I fell on some old tracks, which revived my hopes a little. I slept this evening by a small brook where I collected cherries and haws enough to make a hearty supper. I was obliged to make further encroachments on the legs of my trousers for fresh bandages for my feet.

During the night I was serenaded by music, in which the grumbling bass of the bears was at times drowned by the less pleasing sharps of the wolves. I partially covered my body this night with some pieces of pine bark which I stripped off a sapless tree.

The country through which I dragged my tired limbs on the 24th was thinly wooded. I suffered much from want of water. About sunset I arrived at a small stream, by the side of which I took up my quarters for the night. The dew fell heavily but I was much too fatigued to go in quest of bark to cover me. Even had I been so inclined, the howling of the wolves would have deterred me from making the dangerous attempt.

There must have been an extraordinary nursery of these animals close to the spot. Between the weak, shrill cries of the young, and the more loud and dreadful howling of the old, I never expected to leave the place alive. I could not sleep. My only weapons of defense were a heap of stones and a stick. Ever and anon some more daring than others approached me. I presented the stick at them as if in the act of leveling a gun, upon which they retired, vented a few yells, advanced a little farther, and after surveying me for some time with their sharp fiery eyes, to which the partial glimpses of the moon had imparted additional ferocity, retreated into the wood.

In this state of fearful agitation I passed the night; but as daylight began to break Nature asserted her supremacy, and I fell into a deep sleep, from which, to judge by the sun, I did not awake until between eight and nine o'clock on the morning of the 25th.

My second bandages having been worn out, I was now obliged to bare my knees for fresh ones. After tying them around my feet and taking a copious draft from the adjoining brook, I recommenced my joyless journey. My course was nearly north-northeast. I got no water during the day, nor any of the wild cherries. Some slight traces of men's feet and a few old horse tracks occasionally crossed my path. They proved that human beings sometimes at least visited that part of the country, and for a moment served to cheer my drooping spirits.

About dusk an immense-sized wolf rushed out of a thick copse a short distance from the pathway, planted himself directly before me in a threatening position, and appeared determined to dispute my passage. He was not more than twenty feet from me.

My situation was desperate. As I knew that the least symptom of fear would be the signal for attack, I presented my stick and shouted as loud as my weak voice would permit. He appeared somewhat startled, and retreated a few steps, still keeping his piercing eyes firmly fixed on me. I advanced a little, when he commenced howling in a most appalling manner.

Supposing his intention was to collect a few of his comrades to assist in making an afternoon repast on my half-famished carcass, I redoubled my cries until I had almost lost the power of utterance, at the same time calling out various names, thinking I might make it appear I was not alone.

An old and a young lynx ran close past me but did not stop. The wolf remained about fifteen minutes in the same position. Whether my wild and fearful exclamations deterred any others from joining him, I cannot say. Finding at length my determination not to flinch, and that no assistance was likely to come, he retreated into the wood.

The shades of night were now descending fast, when I came to a verdant spot surrounded by small trees and full of rushes, which induced me to hope for water. A shallow lake or pond had been there, which the long drought and heat had dried up.

I then pulled a quantity of rushes and spread them at the foot of a large stone, which I intended for my pillow. As I was about throwing myself down, a rattlesnake coiled with head erect and the forked tongue extended in a state of frightful oscillation, caught my eye immediately under the stone. I instantly retreated a short distance but, assuming fresh courage, soon dispatched it with my stick. On examining the spot more minutely, a large cluster of them appeared under the stone, the whole of which I rooted out and destroyed. This was hardly accomplished when upwards of a dozen snakes of different descriptions, chiefly dark brown, blue, and green, made their appearance. They were much quicker in their movements than their rattle-tailed brethren; and I could only kill a few of them.

This was a peculiarly soul-trying moment. I had tasted no fruit since the morning before, and after a painful day's march under a burning sun could not procure a drop of water to allay my feverish thirst. I was surrounded by a murderous brood of serpents and ferocious beasts of prey.

Having collected a fresh supply of rushes, which I spread some distance from the spot where I massacred the reptiles, I threw myself on them and was permitted through divine goodness to enjoy a night of undisturbed repose.

I arose on the morning of the 26th considerably refreshed and took a northerly course. Prickly thorns and small sharp stones added greatly to the pain of my tortured feet, and obliged me to make further encroachments on my nether garments for fresh bandages. The want of water now rendered me extremely weak and feverish. I had nearly abandoned all hopes of relief when, about half past four or five o'clock, the old pathway turned from the prairie grounds into a thickly-wooded country. I had not advanced half a mile when I heard a noise resembling a waterfall, to which I hastened my tottering steps, in a few minutes arriving on the banks of a deep and narrow rivulet. I threw myself into the water. Here were plenty of rose hips and cherries on which, with the water, I made a most delicious repast.

On looking about for a place to sleep, I observed lying on the ground the hollow trunk of a large pine, which had been destroyed by lightning. I retreated into the cavity and, having covered myself completely with large pieces of loose bark, quickly fell asleep.

My repose was not of long duration. At the end of about two hours I was awakened by the growling of a bear, which had removed part of the bark and was leaning over me with his snout, hesitating as to the means he should adopt to dislodge me, the narrow limits of the trunk which confined my body preventing him from making the attack with advantage. I instantly sprang up, seized my stick, and uttered a loud cry, which startled him and caused him to recede a few steps, when he stopped and turned about, apparently doubtful whether he would commence an attack. He determined on an assault. Feeling I had not sufficient strength to meet such an unequal enemy, I thought it prudent to retreat and accordingly scrambled up an adjoining tree.

My flight gave fresh impulse to his courage, and he commenced ascending after me. I succeeded, however, in gaining a branch, which gave me a decided advantage over him. I was enabled to annoy his muzzle and claws in such a manner with my stick as effectually to check his progress.

After scraping the bark for some time with rage and disappointment, he gave up the task and retired to my late dormitory, of which he took possession. The fear of falling off, in case I was overcome by sleep, induced me to make several attempts to descend, but each attempt aroused my ursine sentinel. After many ineffectual efforts I was obliged to remain there the rest of the night. I fixed myself in that part of the trunk from which the principal grand branches forked and which prevented me from falling during my fitful slumbers

On the morning of the 27th, a little after sunrise, the bear quitted the trunk, shook himself, and slowly disappeared in search of his morning repast. After waiting some time, apprehensive of his return, I descended and resumed my journey through the woods in a north-northeast direction.

In a few hours all of my anxiety of the preceding night was more than compensated by falling in with a well-beaten horse-path, with fresh traces on it, both of hoofs and human feet. About six in the evening I arrived at a spot where a party must have slept the preceding night. Round the remains of a large fire which was still burning were scattered several half-picked bones of grouse, partridges and ducks, all of which I collected with economical industry. After devouring the flesh I broiled the bones. The whole scarcely sufficed to give me a moderate meal, but yet afforded a most seasonable relief to my famished body. I enjoyed a comfortable sleep this night close to the fire, uninterrupted by any nocturnal visitor.

On the morning of the 28th I set off with cheerful spirits. Late in the evening I arrived at a stagnant pool from which I merely moistened my lips and, having covered myself with some birch bark, slept by its side. The bears and wolves occasionally serenaded me during the night, but I did not see any of them.

I rose early on the morning of the 29th and followed the fresh traces all day. I passed the night by the side of a small stream where I got a sufficient supply of rose hips and cherries. A few distant growls awoke me at intervals but no animal appeared.

On the 30th the path took a more easterly turn, and the woods became thicker and more gloomy. I had now nearly consumed the remnants of my trousers in bandages for my wretched feet and with the exception of my shirt was almost naked.

The horse-tracks every moment appeared more fresh, and fed my hopes. Late in the evening I arrived at a spot where the path branched off in different directions. One led up rather a steep hill, the other descended into a valley, and the tracks of both were equally recent. I took the higher but after proceeding a few hundred paces through a deep wood, I returned, and descended the lower path.

I had not advanced far when I imagined I heard the neighing of a horse. I listened with breathless attention and became convinced it was no illusion. A few paces farther brought me in sight of several of those noble animals sporting in a handsome meadow, from which I was separated by a rapid stream. With some difficulty I crossed over. One of the horses approached me. "His neigh was like the bidding of a monarch and his countenance enforced homage."

Two Indian women perceived me. They instantly led to a hut which appeared at the farthest end of the meadow. This movement made me doubt whether I had arrived among friends or enemies, but my apprehensions were quickly dissipated by the approach of two men, who came running to me in the most friendly manner.

On seeing the lacerated state of my feet, they carried me in their arms to a comfortable dwelling covered with deer-skins. To wash and dress my torn limbs, roast some roots, and boil a small salmon, seemed but the business of a moment.

I collected from their signs that they were aware of my being lost, and that they, with other Indians, had been out several days scouring the woods and plains in search of me. I also understood from them that our party was only a few hours' march from their habitation.

The morning of the 31st was far advanced when I awoke. A considerable stream called

Coeur d'Alene River flowed close to the hut. The old man and his son accompanied me. We crossed the river in a canoe, after which they brought over three horses, and having enveloped my body in an Indian mantle of deer-skin, we mounted.

We arrived in a clear wood in which, with joy unutterable, I observed our Canadians at work hewing timber. I rode between the two natives. One of our men named *Francois Gardepie* joined us on horseback. My deer-skin robe and sun-burnt features completely set his powers of recognition at defiance. He addressed me as an Indian.

I replied in French by asking him how all our people were. Poor Francois appeared electrified and galloped into the wood, vociferating, *"O mes amis! mes amis il est trouve!—Oui, oui, il est trouve!"*

"Qui qui?" asked his comrades.

"Monsieur Cox! Monsieur Cox!" replied Francois.

Away went saws, hatchets, and axes, and each man rushed forward to the tents where we had by this time arrived. The men were allowed a holiday.

5

ASTORIA TO THE BRITISH

THE DIFFERENT PARTIES HAVING ASSEMBLED at Spokan House, we took our departure from that establishment on the 25th of May on our return to Astoria with the produce of our winter's trade. Mr. Pillet was left in charge of the fort with four men. We had twenty-eight loaded horses. On the 30th of May we reached the entrance of the creek off Lewis River, where we had left our barge and canoes.

In the course of this journey we passed some of the places at which I had slept during my wanderings in the preceding August. I pointed out to my fellow-travelers several heaps of stones which I had piled together and on which I had scratched my name.

We were detained a couple of days at the entrance of the creek to repair the barge and canoes in consequence of the Indians having taken a quantity of nails out of the former.

Our tents were pitched close to the village and we kept no watch during the first night.

Our confidence, however, was misplaced, for in the morning we discovered that a daring robbery had been committed. Mr. Clarke immediately assembled the principal Indians; told them of the robbery; declared if the stolen property were returned he would pardon the offender; but added, if it were not and he should find the thief, he would hang him. The chief, with several others, promised they would use their best exertions to discover the delinquent and bring back the property. The day passed without tidings of either.

On the second night two sentinels were placed at each end of the camp with orders to conceal themselves and keep a sharp lookout. Shortly after midnight they observed the figure of a man creeping slowly out of one of the tents. They silently watched his progress until they saw him in the act of jumping into a small canoe which he had in the creek, upon which they sprung forward, stopped the canoe, and seized him.

Most of the property was found in the canoe, but he refused to give an account of the remainder. We had not the slightest suspicion of this man, who had been remarkably well treated by us, in consequence of which and the aggravated nature of the robbery, Mr. Clarke determined to put his threat into execution. He accordingly ordered a temporary gallows to be erected, and had the arms and legs of the culprit pinioned.

About eight o'clock in the morning of the 1st of June he assembled the chief and all the Indians of the village. He made a short speech in which he told them that the prisoner had abused his confidence, violated the rights of hospitality, and committed an offence for which he ought to suffer death. From an anxiety to keep on good terms with all their nation, he had overlooked many thefts committed while he had been there last August. In order to show it was not fear that prevented him from taking an earlier notice of such aggressions, he had now resolved that this robber should be hanged.

The Indians acquiesced in this decision. The chief declared that the prisoner did not belong to their tribe, but was a kind of outlaw of whom they were all afraid.

The gallows being now prepared, Mr. Clarke gave the signal, and after great resistance, during which he screamed in the most frightful manner, the wretched criminal was launched into eternity. His countrymen looked on the whole proceeding with the greatest unconcern, but the unfortunate being himself exhibited none of that wonderful self-command, or stoical indifference to death, which we observed in others and for which Indians in general are so celebrated.

We arrived at Astoria on the 11th of June, 1813, without incurring any material accident.

We found all our friends in good health, but a total revolution had taken place in the affairs of the Company. Messrs. John George M'Tavish and Joseph La Rocque of the North West Company, with two canoes and sixteen men, had arrived a few days before us. From these gentlemen we learned for the first time that war had been declared the year before between Great Britain and the United States, and that in consequence of the strict blockade of the American ports by British cruisers, no vessels would venture to proceed to our remote establishment during the continuation of hostilities.

These unlucky and unexpected circumstances, joined to the impossibility of sustaining ourselves another year in the country without fresh supplies, induced our proprietory to enter into negotiations with Mr. M'Tavish, who had been authorized by the North West Company to treat with them.

In a few weeks an amicable arrangement was made, by which Mr. M'Tavish agreed to purchase all the furs, merchandise, provisions, etc., of our Company at a certain valuation at the same time offering to those who should wish to join the North West Company, and remain in the country, the same terms as if they had originally been members of that company. Messrs. Ross, M'Lennan, and I took advantage of these liberal proposals. The Americans of course preferred returning to their own country.

The pleasure I experienced in joining an establishment every member of which was a fellow-subject, was mingled with deep regret at parting from so many of my late associates.

6

CURE FOR RHEUMATISM

Cox was kept busy by his new bosses. From July 5 to October 11, 1813, he traveled two thousand three hundred miles. Then on October 29 he was sent back upriver to the Flathead country.

WE REACHED MR. M'MILLAN on the 24th of December. The fort had a good trading store, a comfortable house for the men, and a snug box for ourselves, all situated on a point formed by the junction of a bold mountain torrent with the Flathead River, and surrounded on all sides with high and thickly wooded hills covered with pine, spruce, larch, beech, birch, and ceder. A large band of the Flathead warriors were encamped about the fort, recently returned from the buffalo country.

M'Millan's tobacco and stock of trading goods had been entirely expended previous to my arrival, and the Indians were much in want of ammunition, etc. My appearance, or I should rather say the goods I brought with me, was therefore a source of joy to both parties. The natives smoked the much-loved weed for several days successively. Our hunters killed a few mountain sheep, and I brought up a bag of flour, a bag of rice, plenty of tea and coffee, some arrowroot, and fifteen gallons of prime rum. We spent a comparatively happy Christmas and, by the side of a blazing fire in a warm room, forgot the sufferings we endured in our dreary progress through the woods.

Very little snow fell after Christmas but the cold was intense, with a clear atmosphere.

I experienced some acute rheumatic attacks in the shoulders and knees, from which I suffered much annoyance. An old Indian proposed to relieve me, provided I consented to follow the mode of cure practiced by him in similar cases on the young warriors of the tribe.

On inquiring the method he intended to pursue, he replied that it merely consisted in getting up early every morning for some weeks, and plunging into the river, and to leave the rest to him. This was a most chilling proposition, for the river was firmly frozen, and an opening had to be made in the ice preparatory to each immersion.

I asked him, "Would it not answer equally well to have the water brought to my bedroom?"

He shook his head and replied he was surprised that a young white chief, who ought to be wise, should ask so foolish a question.

On reflecting, however, that rheumatism was a stranger among Indians, while numbers of our people were martyrs to it, and above all that I was upwards of three thousand miles from any professional assistance, I determined to adopt the disagreeable expedient, and commenced operations the following morning.

The Indian first broke a hole in the ice sufficiently large enough to admit us both, upon which he made a signal that all was ready. Enveloped in a large buffalo robe, I proceeded to the spot and, throwing off my covering, we both jumped into the frigid orifice together. He immediately commenced rubbing my shoulders, back, and loins. My hair in the meantime became ornamented with icicles; and while the lower joints were undergoing their friction, my face, neck, and shoulders were incased in a thin covering of ice.

On getting released, I rolled a blanket about me and ran back to the bedroom, in which I had previously ordered a good fire. In a few minutes I experienced a warm glow all over my body.

Chilling and disagreeable as these matinal

ablutions were, yet, as I found them so beneficial, I continued them for twenty-five days, at the expiration of which my physician was pleased to say that no more were necessary, and that I had done my duty like a wise man. I was never after troubled with a rheumatic pain!

7

JANE BARNES

[THE NAME OF FORT ASTORIA WAS CHANGED to that of Fort George. Here on April 22, 1814, arrived the long-looked-for *Isaac Todd*. One of the passengers was another M'Tavish whose first name was Donald, with authority to act as governor of the North West Company. He brought with him a flaxen-haired, blue-eyed beauty who has lived in history as the first white woman on the Columbia River.]

Miss Jane Barnes had been a lively barmaid at a hotel in Portsmouth at which Mr. Mac had stopped preparatory to his embarkation. This gentleman, being rather of an amorous temperament, proposed the trip to Miss Jane, who, nothing loath, threw herself on his protection regardless of consequences.

After encountering the perils of a long sea voyage she found herself an object of interest to the residents of the fort, and the greatest curiosity that ever gratified the wandering eyes of the blubber-loving aboriginals of the northwest coast of America. The Indians daily thronged in numbers to our fort for the mere purpose of gazing on, and admiring the fair beauty, every article of whose dress was examined with the most minute scrutiny.

She had rather an extravagant wardrobe, and each day exhibited her in a new dress, which she always managed in a manner to display her figure to the best advantage. One day, her head, decorated with feathers and flowers produced the greatest surprise; the next, her hair, braided and unconcealed by any covering, excited equal wonder and admiration. The young women felt almost afraid to approach her, and the old were highly gratified at being permitted to touch her person.

Some of the chiefs having learned that her protector intended to send her home, thought to prevent such a measure by making proposals of marriage. One of them in particular, the son of Comcomly, the principal chief of the Chinooks, came to the fort attired in his richest dress, his face fancifully bedaubed with red paint, and his body redolent of whale oil. He was young and had four native wives. He told her that if she would become his wife he would send one hundred sea-otters to her relations. He would never ask her to carry wood, draw water, dig for roots, or hunt provisions. He would make her mistress of his other wives and permit her to sit at her ease from morning to night and wear her own clothes. She should always have abundance of fat salmon, anchovies, and elk, and be allowed to smoke as many pipes of tobacco during the day as she thought proper.

Together with many other flattering inducements, the tithe of which would have shaken the constancy of a score of the chastest brown vestals that ever flourished among the lower tribes of the Columbia.

These tempting offers, however, had no charms for Jane. Her long voyage had not yet eradicated certain Anglican predilections respecting mankind, which she had contracted in the country of her birth, and among which she did not include a flat head, a half naked body, or a copper-colored skin besmeared with whale oil.

Her native inamorato made several other ineffectual proposals, but finding her inflexible he declared he would never come near the fort while she remained there. We shortly learned afterwards that he had concerted a plan with some daring young men of his tribe to carry her off while she was walking on the beach (her general custom every evening while the gentlemen were at dinner), a practice which, after this information, she was obliged to discontinue.

Miss Barnes was fond of quotations. One of the clerks was one day defending the native and half-breed women whose characters she had violently attacked, and he recriminated in no very measured language on the conduct of the white ladies.

"O Mr. Mac," said she, "I suppose you agree with Shakespeare that 'every woman is at heart a rake.' "

"Pope, ma'am, if you please."

"Pope! Pope!" replied Jane. "Bless me, sir, you must be wrong; *rake* is certainly the word —I never heard of but one female Pope."

Then in order to terminate the argument, she pretended to read an old newspaper which she held in her hand. He quickly discovered by her keeping the wrong end uppermost that she did not know a syllable of its contents. He quitted her abruptly. As he was coming out I met him at the door, a wicked and malicious grin ruffling his sunburnt features.

"Well, Mac," said I, "what's the matter? You seem annoyed."

"What do you think," he replied, "I have just had a conversation with that fine-looking damsel there, who looks down with such contempt on our women, and may I be d----d if the b---h understands B from a buffalo!"

Her supposed education was the only excuse in his opinion to justify her usurpation of superiority—that gone, he judged her poor indeed.

Mr. Mac at first intended to have brought her with him across the continent to Montreal; but on learning the impracticability of her performing such an arduous journey, he abandoned that idea and made arrangements with the captain for her return to England by way of Canton.

A few words more and I shall have done with Miss Barnes. On the arrival of the vessel at Canton she became an object of curiosity and admiration among the inhabitants of the Celestial empire. An English gentleman of great wealth connected with the East-India Company, offered her a splendid establishment. It was infinitely superior to any of the proposals made by the Chinook nobility, and far beyond anything she could ever expect in England. It was therefore prudently accepted, and the last account I heard of her stated that she was then enjoying all the luxuries of eastern magnificence.

8

CHINOOKS, CLATSOPS, AND NEIGHBORS

1814. WE REMAINED a couple of months this summer at Fort George, making the necessary arrangements for our winter's campaign. During this period we made several excursions on pleasure or business to the villages of the various tribes from one to three days' journey from the fort.

They differ little from each other in laws, manners, or customs. Were I to make a distinction, I would say the Cathlamahs are the most tranquil, the Killymucks the most roguish, the Clatsops the most honest, and the Chinooks the most incontinent.

The abominable custom of flattening their heads prevails among them all. Immediately after birth the infant is placed in a kind of oblong cradle formed like a trough, with moss under it. One end, on which the head reposes, is more elevated than the rest. A padding is then placed on the forehead with a piece of cedar bark over it. By means of cords passed through small holes on each side of the cradle, the padding is pressed against the head. It is kept in this manner upwards of a year, and is not I believe attended with much pain.

The appearance of the infant, however, while in this state of compression is frightful. Its little black eyes, forced out by the tightness of the bandages, resemble those of a mouse choked in a trap. When released from this inhuman process, the head is perfectly flattened, and the upper part of it seldom exceeds an inch in thickness. It never afterwards recovers its rotundity. They deem it an essential point of beauty. Doctor Swan, on examining the skulls I had taken, candidly confessed that nothing short of ocular demonstration could have convinced him of the possibility of moulding the human head into such a form.

They allege as an excuse for this custom that all their slaves have round heads. Accordingly every child of a bondsman, who is not adopted by the tribe, inherits not only his father's degradation but his parental rotundity of cranium.

This deformity is unredeemed by any peculiar beauty either in features or person.

The height of the men varies from five feet to five feet six. The nose is rather flat, with distended nostrils. The limbs of the men are in general well shaped. The women, owing to the tight ligatures which they wear on the lower part of the legs, are quite bandy, with thick ankles, and broad flat feet. They have loose hanging breasts, slit ears, and perforated noses, which, added to greasy heads and bodies saturated with fish-oil, constitute the sum total of their personal attractions.

The good qualities of these Indians are few; their vices many. In the latter may be classed thieving, lying, incontinence, gambling, and cruelty. They are also perfect hypocrites. Even the natives of the same village, while they feign an outward appearance of friendship, indulge in a certain propensity called backbiting.

Their bravery is rather doubtful but what they want in courage they make up in effrontery.I have seen a fellow stopped on suspicion of stealing an axe. He denied the charge with the most barefaced impudence. When the stolen article was pulled from under his robe, instead of expressing any regret he burst out laughing and alleged he was only joking. One of the men gave him a few kicks, which he endured with great *sang-*

froid. On joining his companions, they received him with smiling countenances and bantered him on the failure of his attempt. They seldom make any resistance to these summary punishments. If the chastisement takes place in the presence of a chief, he seems delighted at the infliction.

They purchase slaves from the neighboring tribes for beaver, otter, beads, etc. I could never learn whether any were taken by them in war. While in good health and able to work, they are well treated, but the moment they fall sick, or become unfit for labor, the unfortunate slaves are totally neglected, and left to perish in the most miserable manner. After death their bodies are thrown without any ceremony at the trunk of a tree or into an adjoining wood. It sometimes happens that a slave is adopted by a family; in which case he is permitted to marry one of the tribe, and his children, by undergoing the flattening process, melt down into the great mass of the community.

About thirty years before this period the smallpox had committed dreadful ravages among these Indians, the vestiges of which were still visible on the countenances of the elderly men and women. The western tribes still remember it with a superstitious dread, of which Mr. M'Dougall took advantage when he learned that the *Tonquin* had been cut off. He assembled several of the chieftains and, showing them a small bottle, declared that it contained the smallpox. Mr. M'Dougall promised that if the white people were not attacked or robbed for the future, the fatal bottle should not be uncorked. He was greatly dreaded by the Indians, who were fully impressed with the idea that he held their fate in his hands. They called him by way of preeminence "the great smallpox chief."

An Indian belonging to a small tribe on the coast to the southward of the Clatsops, occasionally visited the fort. His history was rather curious. His skin was fair, his face partially freckled, and his hair quite red. He was about five feet ten inches high, was slender, but remarkably well made. His head had not undergone the flattening process. He was called Jack Ramsay in consequence of that name having been punctured on his left arm.

The Indians allege that his father was an English sailor who had deserted from a sailing vessel, and had lived many years among their tribe, one of whom he married; that when Jack was born he insisted on preserving the child's head in its natural state, and while young had punctured the arm in the above manner. Old Ramsay had died about twenty years before this period. He had several more children but Jack was the only red-headed one among them. He was the only half-breed I ever saw with red hair, as that race in general partake of the swarthy hue derived from the maternal ancestors.

Poor Jack was fond of his father's countrymen, and had the decency to wear trousers whenever he came to the fort. We therefore made a collection of old clothes for his use, sufficient to last him for many years.

The ideas of these Indians on the subject of a future state do not differ much from the opinions entertained by the natives of the interior. They believe that those who have not committed murder; have fulfilled the relative duties of son, father, and husband; have been good fishermen, etc., will after their death go to a place of happiness in which they will find an abundant supply of fish, fruit, etc. Those who have followed a contrary course of life will be condemned to a cold and barren country in which bitter fruits and salt water will form their principal means of subsistence.

Civilized countries are not exempt from superstition; it is therefore not surprising to find it existing among untutored savages. They believe that if salmon be cut crossways the fishery will be unproductive and that a famine will follow. In the summer of 1811 they at first brought but a small quantity to the people who were then building the fort. As Mr. M'Dougall knew there was no scarcity, he reproached the chiefs for furnishing such a scant supply. They admitted the charge but assigned as a reason their fears that the white people would cut it in the unlucky way. Mr. M'Dougall promised to follow their plan, upon which they brought a tolerable good quantity but all roasted; and which, in order to avoid displeasing them, our people were obliged to eat before sunset each day.

A man may have as many wives as his means will permit him to keep. They live together in the greatest harmony. Although their lord may love one more than another, it causes no jealousy or disunion among the rest.

Many of these women who have followed a depraved course of life before marriage, become excellent and faithful wives afterwards. In the early part of this summer, one of the clerks who had been out on a trading excursion, happened to be present at a marriage in a Clatsop village. He was surprised

ape Horn, 25 miles upriver from Vancouver, where the Evergreen Highway passes under a snow
ed. This photograph, made by B. A. Gifford in 1909, shows it still the same as Cox saw it as
the French-Canadian voyageurs paddled him by.

Cox said, "The Columbia is a noble river, uninterrupted by rapids for 170 miles . . . seldom less than a mile wide."

Cox came to the foot of Celilo Falls on July 4, 1812, and camped. "We only made one-half of the portage the first day."

Rooster Rock, another of the big stone exhibits along the Columbia, as photographed by E. H. Moorehouse. Cox gave few details of the scenery as such; he was an action writer rather than a descriptive one.

Cabbage Rock, four miles north of The Dalles, in a view by Major Lee Moorehouse, noted Pendleton photographer. The stony mass so neatly balanced on its slender stem, was passed up by Cox as literary material.

at recognizing in the bride an old *chere amie* who in the preceding year had spent three weeks with him in his tent, actually decorated with some of the baubles he had then given her. His eye caught hers for a moment but his appearance excited not the least emotion, and she passed him by as one whom she had never seen. A few days afterwards she came to the fort accompanied by her husband and other Indians. She remained at the gate while the men were selling some fish in the trading store. Her old lover, observing her alone, attempted to renew their former acquaintance; but she betrayed no symptom of recognition, and in a cold distant manner told him to go about his business.

All the Indians of the Columbia entertain a strong aversion to ardent spirits, which they regard as poison. They allege that slaves only drink to excess and that drunkenness is degrading to free men. On one occasion some of the gentlemen at Fort George induced a son of Comcomly the chief to drink a few glasses of rum. Intoxication quickly followed, accompanied by sickness; in which condition he returned home to his father's house, and for a couple of days remained in a state of stupor. The old chief subsequently reproached the people at the fort for having degraded his son by making him drunk and thereby exposing him to the laughter of the slaves.

Each village is governed by its own chief. He possesses little authority, and is respected in proportion to the number of wives, slaves, &c. which he may keep. The greater number of these, the greater the chief. He is entitled, however, to considerable posthumous honor. At his death the tribe go into mourning by cutting their hair, and for some months continue to chant a kind of funeral dirge to his memory.

The great mass of the American Indians, in their warlike encounters, fall suddenly on their enemies. The plan adopted by the Chinooks forms an honorable exception to this system. They gave notice to the enemy of the day on which they intend to make the attack. Having previously engaged as auxiliaries a number of young men whom they pay for that purpose, they embark in their canoes for the scene of action.

On arriving at the enemy's village they enter into a parley, and endeavor by negotiation to terminate the quarrel amicably. Sometimes a third party, who preserves a strict neutrality, undertakes the office of mediator. Should their joint efforts fail in procuring redress, they immediately prepare for action. Should the day be far advanced, the combat is deferred by mutual consent till the following morning. They pass the intervening night in frightful yells, and making abusive and insulting language to each other.

They generally fight from their canoes, which they take care to incline to one side, presenting the higher flank to the enemy. In this position, with their bodies quite bent, the battle commences. Owing to the cover of their canoes and their impenetrable armor, it is seldom bloody. As soon as one or two men fall, the party to whom they belonged acknowledge themselves vanquished and the combat ceases. If the assailants be unsuccessful, they return without redress. If conquerors, they receive various presents from the vanquished party in addition to their original demand.

Their culinary articles consist of a large square kettle made of cedar wood, a few platters made of ash, and awkward spoons made of the same material. Their mode of cooking, however, is more expeditious than ours. Having put a certain quantity of water into a kettle, they throw in several hot stones, which quickly causes the water to boil. The fish or meat is then put in, and the steam is kept from evaporating by a small mat thrown over the kettle. By this system a large salmon will be boiled in less than twenty minutes.

They are not scrupulously clean in their cooking. A kettle in which salmon is boiled in the morning may have elk dressed in the same evening, and the following day be doomed to cook a dish of sturgeon, without being washed out or scarcely rinsed.

In felling the timber for their houses, and in the laborious operation of forming their canoes, they had not, previous to our arrival, an axe. Their only instruments consisted of a chisel generally formed out of an old file, a kind of oblong stone which they used as a hammer, and a mallet made of spruce knot, well oiled and hardened by fire. With these wretched tools they cut down trees from thirty to forty feet in circumference; and with unparalleled patience and perseverence continued their tedious and laborious undertaking until their domicile was roofed or their canoe fit to encounter the turbulent waves of the Columbia.

Gambling is one of their most incorrigible vices. So inveterately are they attached to it that the unfortunate gamester often finds himself stripped of slaves, beads, *haiqua,* and even nets. Their common game is a simple kind of hazard. One man takes a small stone which he changes for some time from hand to hand, all the while humming a slow monotonous air. The bet is then made. According as his adversary succeeds in guessing the hand in which the stone is concealed, he wins or loses. They seldom cheat, and submit to their losses with the most philosophical resignation.

Every village has its quack doctor; or, as they call him, "the strong man of medicine." The moment the native is attacked with sickness, no matter of what description, the physician is sent for. He immediately commences operations by stretching his patient on his back, while a number of his friends and relatives surround him, each carrying a long and a short stick, with which they beat time to a mournful air which the doctor chants and in which they join at intervals. Sometimes a slave is dispatched to the roof of the house, which he belabors most energetically with his drumsticks, joining at the same time with a loud voice the chorus inside. The man of medicine then kneels, and presses with all his force his two fists on the patient's stomach. The unfortunate man, tortured by the pain produced by this violent operation, utters the most piercing cries; but his voice is drowned by the doctor and the bystanders, who chant loud and louder still the mighty "song of medicine."

At the end of each stanza the operator seizes the patient's hands, which he joins together and blows on. He thus continues alternately pressing and blowing until a small white stone, which he had previously placed in the patient's mouth, is forced out. This he exhibits with a triumphant air to the man's relations; and, with all the confidence and pomposity of modern quackery, assures them the disease is destroyed and that the patient must undoubtedly will recover.

It frequently happens that a man who might have been cured by a simple dose of medicine, is by this abominable system destroyed. Whether recovery or death be the consequence, the quack is equally recompensed. Some of the more intelligent undoubtedly perceive the imposition which these fellows practice; but the great faith which the ignorant and superstitious multitude have in their skill, deters any man from exposing their quackery. Latterly, however, numbers of their sick have applied for relief and assistance at Fort George. As our prescriptions have been generally attended with success, their belief in the infallibility of those jugglers has been considerably weakened.

When a Chinook dies, his remains are deposited in a small canoe, the body being previously enveloped in skins or mats. His bow, arrows, and other articles are laid by his side. The canoe is then placed on a high platform near the river's side, or on rocks out of the reach of the tide, and other mats tied over it. If the relations of the deceased can afford it, they place a larger canoe reversed over the one containing his body, and both are firmly tied together. His wives, relatives, and slaves go into mourning by cutting their hair; and for some time after his death repair twice a day, at the rising and setting of the sun, to an adjoining wood to chant his funeral dirge.

9

DUEL AT SPOKAN HOUSE

On the 5th of August, 1814, we left Fort George. We met a few of the Nez Perces at the mouth of the Lewis River. They appeared friendly and sold us some horses. From this place nothing particular occured until the 23rd of August, on which day we arrived at Oakinagan. The various parties separated for their summer destinations. Mine was Spokan house, in company with Messrs Stewart, M'Millan, and M'Donald.

While we were here a curious incident occured between Mr. M'Donald and an Indian.

At the period I speak of M'Donald had been ten years absent from Canada and had traveled over an immense extent of Indian country. He seldom remained more than one winter at any particular place, and had a greater facility of acquiring than of retaining the language of the various tribes with whom he came in contact. He was subject to temporary fits of abstraction, during which the country of his auditory was forgotten, and their lingual knowledge set at defiance by the most strange and ludicrous *melange* of Gaelic, English, French, and half a dozen Indian dialects.

Whenever anything occurred to ruffle his temper, it was highly amusing to hear him give vent to his passions in *Diaouls, God d--s, Sacres,* and invocations to the evil spirit in Indian. He was, however, a good-natured, inoffensive companion, easily irritated and as easily appeased.

His appearance was very striking. In height he was six feet four inches, with broad shoulders, large bushy whiskers, and red hair, which for some years had not felt the scissors and which, sometimes falling over his face and shoulders, gave to his contenance a wild and uncouth appearance. He had taken a Spokan wife, by whom he had two children. A great portion of his leisure time was spent in the company of her relations, by whom and indeed by the Indians in general he was highly beloved. Their affection, however, was chastened by a moderate degree of fear, with which his gigantic body and indomitable bravery inspired them.

One day, as we were sitting down to dinner, one of our men, followed by a native, rushed into the dining room and requested we would instantly repair to the village to prevent bloodshed, as M'Donald was about to fight a duel with one of the chiefs.

We ran to the scene of action and found our friend surrounded by a number of Indians, all of whom kept at a respectable distance. He had his fowling-piece, which he changed from one hand to the other, and appeared violently chafed. The chief stood about twenty yards from him, and the following colloquy took place between them:

M'Donald—Come on now, you rascal, you toad, you dog! Will you fight.

Indian—I will—but you are a foolish man. A chief should not be passionate. I always thought the white chiefs were wise men.

M'Donald—I want none of your jaw. I say you cheated me. You're a dog! *Will* you fight?

Indian— You are not wise. You get angry like a woman. But I will fight. Let us go to the wood. Are you ready?

M'Donald—Why, you d - - d rascal, what do you mean? I'll fight you here. Take your distance like a brave man, face to face, and we'll draw lots for the first shot, or fire together, whichever you please.

INDIAN—You are a greater fool than I thought you were. Who ever heard of a wise warrior standing before his enemy's gun to be shot like a dog? No one but a fool of white man would do so.

M'DONALD—What do you mean? What *way* do you want to fight?

INDIAN—The way that all red warriors fight. let us take our guns and retire to yonder wood. Place yourself behind one tree. I will take my stand behind another. Then we shall see who will shoot the other first.

M'DONALD—You are afraid and you're a coward.

INDIAN—I am not afraid and you are a fool.

M'DONALD—Come then, d - - n my eyes if I care. Here's at you your own way.

And he was about proceeding to the woods when we interfered, had the combatants disarmed, and after much entreaty induced our brave to return to thc fort.

The quarrel originated in a gambling transaction. M'Donald imagined he had been cheated and under that impression struck the chief and called him a rogue. The latter told him he took advantage of his size and strength and would not meet him on equal terms with his gun. This imputation roused all M'Donald's ire. He instantly darted into the field with his fowling-piece, followed by the chief. By our arrival we prevented an encounter which in all probability would have proved fatal to our friend.

The gigantic figure, long red flowing locks, foaming mouth, and violent gesticulation of M'Donald, presented a striking and characteristic contrast to the calm and immutable features of the chieftain. M'Donald's proposition appeared to him so much at variance with his received notions of wisdom that he could not comprehend how any man in his senses could make such an offer.

On explaining to the chief afterwards the *civilized* mode of deciding gentlemanly quarrels, he manifested the utmost incredulity and declared that he could not conceive how people so wise in other respects should be guilty of such foolishness. When we assured him in the most positive manner that we were stating the facts, he shook his head and said, "I see plainly there are fools everywhere".

10

EIGHT CANOES AND A BATTLE

ON THE 24TH OF OCTOBER we proceeded overland with the produce of the summer's trade to Oakinagan, where, being joined by the people of that district, we embarked for Fort George, at which place we arrived on the 8th of November. We remained only a few days at Fort George, from which place we took our departure for the interior on the 18th of November.

On arriving a few miles above the entrance of the Wallah Wallah river a number of canoes filled with natives paddled down on our brigade, apparently without any hostile design. We were on the south side advancing slowly with the poles.

We had eight canoes. Our party consisted of Messrs Keith, Stewart, LaRocque, McTavish, McDonald, McKay, McKenzie, Montour, and myself. We had fifty-four canoemen, including six Sandwich islanders.

The Indians at first asked a little tobacco

from Mr. Keith, which he gave them. They then proceeded to Mr. Stewart, who also gave them a small quanitity. After which they dropped down on Messrs LaRocque and McMillan, from whose canoe they attempted to take some goods by force but were repulsed by the men, who struck their hands with the paddles. They next came to McDonald and seized a bale of tobacco which was in the forepart of the canoe and which they attempted to take out. At the same time my canoe was stopped, as well as those in the rear. A determined resolution was evinced to plunder us by force.

We were awkwardly circumstanced. The only arms at hand were those in the possession of the officers. With the exception of the paddles, the men had no weapons ready.

Anxious to avoid coming to extremities as long as possible, we endeavored to keep them in check with the paddles. Mr Keith gave orders not to fire while there was a possibility of saving the property.

The fellow who had seized the bale in McDonald's canoe was a tall athletic man. He resisted all their entreaties to let it go, and had partly taken it out of the canoe when McKay gave him a severe blow with the butt of his gun, which obliged him to drop the prize. He instantly placed an arrow in his bow, which he presented at McDonald. The latter coolly stretched forth his brawny arm, seized the arrow, which he broke and threw into the fellow's face. The savage, enraged at being thus foiled, ordered his canoe to push off, and was just in the act of letting fly another arrow when McKay fired and hit him in the forehead. He instantly fell, upon which two of his companions bent their bows. But before their arrows had time to wing their flight, McDonald's double-barreled gun stopped them. He shot one between the eyes. The ball from the second barrel lodged in the shoulder of the survivor.

The moment they fell, a shower of arrows was discharged at us. But owing to the undulating motion of their canoes, as well as ours, we escaped uninjured. Orders were now issued to such as had their arms ready, to fire; but in a moment our assailants became invisible. After they had discharged their arrows, they had thrown themselves prostrate in their canoes, which, drifting rapidly down the current, were quickly carried beyond the reach of our shot.

We lost no time in putting ashore for the purpose of arming the men and distributing ammunition. The Columbia at this place was nearly a mile wide. Night was fast approaching.

A short distance higher up in the center of the river lay a narrow island, about two miles in length, quite low, void of timber, and covered with small stones and sand. It was deemed the safest place to withstand an attack or preventing a surprise. Orders were therefore given to collect as much driftwood as possible on the main shore for the purpose of cooking. This was speedily effected, after which we pushed off; but had not proceeded more than one hundred yards when several arrows were discharged at us from the side we had just left, although at the time we embarked no Indian was visible for miles around.

The brigade was divided into three watches. The night was dark, cold, and stormy, with occasional showers of rain. It was judged prudent to extinguish the campfires lest their light might serve as a beacon to the Indians in attacking us. This precaution, although by no means relished by the men, probably saved the party, for about an hour before daybreak several of the savages were discovered close to camp, silently approaching on their hands and feet. On being fired at by our sentinels, they quickly retreated. Shortly after we heard the sound of their paddles quitting the island.

Shortly after daybreak a council of war was held. After some discussion, we determined to quit the island, demand a parley, and offer a quantity of goods to appease the relations of the deceased.

It blew a strong gale during the day, which prevented us from embarking and constrained us to pass another melancholy night on the island, without wood sufficient to make a solitary fire.

Towards midnight the storm subsided. Mr. Keith commanded the second watch. I was sitting with him at the extremity of the camp when we observed a large fire on a hill in a northwest direction. It was immediately answered by one in the opposite point, followed by others to the eastward and westward. The indistinct sounds of paddles from canoes crossing and recrossing, afforded strong proofs that our enemies, by vigilant watching and constant communication, had determined that we should not escape them in the dark.

Shortly after these threatening indications, a flight of ravens passed quietly over our heads, the fluttering of whose wings were scarcely audible. Some of the Canadians were near us. One of them named Landreville in rather a dejected tone said to his comrades, "My friends, it is useless to hope. Our doom is fixed. Tomorrow we shall die."

"*Cher frere,* what do you mean?" eagerly inquired half a dozen voices.

"Behold yon ravens", he replied. "Their appearance by night in times of danger betokens approaching death. I cannot be mistaken. They know our fate and will hover about us until the arrows of the savages give them a banquet on our blood."

Landreville in other respects was a steady, sensible man, but, like his countrymen, deeply imbued with superstitious ideas. Mr. Keith saw the bad impression which these ominous forebodings were likely to produce on the men, and at once determined to counteract it.

This he knew it would have been useless to attempt by reasoning with people whose minds such absurd notions would have closed against conviction, and therefore thought it better to combat their prejudices with their own weapons.

"I have no doubt, my friends", said he, "that the appearance of ravens at night portends either death or some great disaster. At the same time I must tell you that no fatality is ever apprehended except their appearance is accompanied by croaking. *Then* indeed the most direful consequences are likely to follow. But when their flight is calm and tranquil, as we have just witnessed, they are always the harbingers of good news."

The poor fellows exclaimed, "You are right, sir, you are right. We believe you, sir. You speak reason. Courage, friends; there's no danger."

The morning of the 1st of December rose cold and bright over the plains of the Columbia as we prepared to quit our cheerless encampment.

After having examined their muskets, and given each man an additional glass of rum, we embarked and in a few minutes reached northern shore, where we landed. None of the natives were visible. We remained about half an hour undecided as to what course we should adopt, when a few mounted Indians made their appearance at some distance. Michel, the interpreter, was sent forward alone, carrying a long pole to which was attached a white hankerchief, and hailed them several times without obtaining an answer.

After a little hesitation two of them approached and demanded to know what we had to say. Michel replied that the white chiefs were anxious to see their chiefs and elders and to have a talk with them on the late disagreeable affair. One of them replied that he would inform his friends, upon which he and his companion galloped off.

In less than half an hour a number of mounted Indians appeared, preceded by about one hundred and fifty warriors on foot, all well armed with guns, spears, tomahawks, bows and well-furnished quivers. They stopped within about fifty yards of our party. Among them we recognized several of the Wallah Wallahs but in vain looked for our old friend Tom Tappam, their chief.

A group of between thirty and forty now approached from the interior. This party consisted of the immediate relatives of the deceased. As they advanced they chanted a death song:

"Rest, brothers, rest! You will be avenged. The tears of your widows shall cease to flow when they behold the blood of your murderers, and your young children shall leap and sing with joy on seeing their scalps. Rest, brothers, in peace, we shall have blood."

Messrs Keith, Stewart, La Rocque, and the interpreter, at length advanced about midway both parties unarmed and demanded to speak with them. Mr. Keith offered them the calumet of peace which they refused to accept in a manner at once cold and repulsive.

Michel was thereupon ordered to tell them that we regretted much that the late unfortunate circumstances had occurred to disturb our friendly intercourse. We were now willing to compensate the relations of the deceased for the loss they had sustained.

They inquired what kind of compensation was intended. On being informed that it consisted of two suits of chiefs' clothes, with blankets, tobacco, and ornaments for the women, it was indignantly refused. Their spokesman stated that no discussion could be entered into until two white men (one of whom should be the big red-headed chief) were delivered to them.

Every eye turned on M'Donald, who, on hearing the demand, "grinned horribly a ghastly smile" and who, but for our interposition, would on the spot have chastized the insolence of the speaker. The men were horrified until Mr. Keith promptly restored their confidence by telling them that such an ignominious demand should never be complied with.

He then addressed the Indians in a calm,

firm voice, and told them that no consideration whatever should induce him to deliver a white man to their vengeance. They had been the original aggressors. In their unjustifiable attempt to seize by force our property the deceased had lost their lives. He was willing to believe the attack was unpremeditated, and under that impression he had made the offer of compensation. He reminded them of our superiority in arms and ammunition. For every man belonging to our party who might fall, ten of their friends at least would suffer. He concluded by requesting them calmly to weigh and consider all these matters. Upon the result of their deliberation would in a great measure depend whether white men would remain in their country or quit it forever.

A violent debate took place among the principal natives. One party advised the demand for the two white men to be withdrawn and to ask in their place a greater quantity of goods and ammunition. The other, by far the most numerous, opposed all compromise, accompanied by the delivery of the victims.

Michel told Mr. Keith that he was afraid an accommodation was impossible. Orders were thereupon issued to prepare for action.

From their motions it became evident that their intention was to outflank us. Messrs Keith and Stewart, who had now abandoned all hope of an amicable termination, called for their arms.

An awful pause ensued when our attention was arrested by the loud tramping of horses. Immediately after twelve mounted warriors dashed into the space between the two parties, where they halted and dismounted. They were headed by a young chief of fine figure who instantly ran up to Mr. Keith, to whom he presented his hand in the most friendly manner. He then commanded our enemies to quit their places of concealment and appear before him. His orders were promptly obeyed.

Having made himself acquainted with the circumstances that led to the deaths of the two Indians, he addressed them in a speech of considerable length.

Through this chief's mediation the various claimants were in a short time fully satisfied. The chieftain whose timely arrival had rescued us from impending destruction was called Morning Star. His age did not exceed twenty-five years. Nineteen scalps decorated the neck of his war horse, the owners of which had all been killed in battle.

We gave the man who had been wounded in the shoulder a chief's coat. To the relations of the men who were killed we gave two coats, two blankets, two fathoms of cloth, two spears, forty bullets and powder, with a quantity of trinkets and two small kettles for their widows. We also distributed nearly half a bale of tobacco among all present. Our youthful deliverer was presented by Mr. Keith with a handsome fowling-piece, and some other valuable articles.

11

A SQUAW MAN'S WRONG SQUAW

I REMAINED at Spokan in company with Messrs Stewart and M'Tavish and passed rather an agreeable winter.

In the course of the winter an incident occurred which threatened at the time to interrupt the harmony that had previously existed between our people and the Spokan Indians.

One of our young clerks, having become tired of celibacy, resolved to take a wife. As none of the Columbia half-breeds had attained a sufficiently mature age, he was necessitated to make his selection from the Spokan tribe. He therefore requested the interpreter to make an inquiry in the village, and ascertain whether any unappropriated comely young woman was willing to become the partner of a juvenile chief.

A pretty-looking damsel, about seventeen years of age, immediately became a candidate for the prize. As her father had died some years before, she was under the guardianship of her mother, who, with her brother, settled the terms of the negotiation. Blankets and kettles were presented to her principal relations, while beads, hawk-bells, &c. were distributed among the remaining kindred.

About nine o'clock at night the bride was conducted to the fort-gate by her mother. After an apathetic parting, she was consigned to the care of one of the men's wives called "the scourer" who had her head and body thoroughly cleansed from all Indian paint and grease with which they had been saturated. After this purification she was handed over to the dressmaker, who instantly discharged her leathern chemise and supplied its place by more appropriate clothing. The following morning, when she appeared in her new habiliments, we thought her one of the most engaging females that we had previously seen of the Spokan nation.

Matters rolled on pleasantly enough for a few days. The youthful couple appeared mutually enamored of each other. But a "litte week" had scarcely passed over their heads when one day about two o'clock a number of young warriors well mounted galloped into the courtyard of the fort armed at all points. Their appearance was so unusual and unlike the general manner of the Spokan nation that we were at a loss to account for it. Vague suspicions of treachery began to flit across our imaginations, but the mystery was shortly cleared up.

The bride, on perceiving the foremost horseman of the band enter the court, instantly fled into an adjoining store, in which she concealed herself. The leader of the band and his associates dismounted and demanded to speak with the principal white chief, at the same time requesting the other chiefs would also appear. His wishes having been complied with, he addressed us in substance to the following effect:

"Three snows have passed away since the white men came from their country to live among the Spokans. When their own hunters could not find an animal, did the Spokans take advantage of their afflictions? Did they rob them of their horses like Sinapoil dogs? Did they say, The white men are now poor and starving and we can easily take all their goods from them? No, we never spoke or even thought of such bad things. The white men came amongst us with confidence and our hearts were glad to see them; they paid us for our fish, for our meat, and for our furs. We thought they were all good people, in particular their chiefs, but I find we were wrong in so thinking."

Here he paused for a short period, after which he thus recommended:

"My relations and myself left our village some days ago for the purpose of hunting. We returned home this morning. Their wives and their children leaped with joy to meet them, and all their hearts were glad but mine. I went to my hut and called on my wife to come forth, but she did not appear. I was sorrowful and hungry and went into my brother's hut where I was told that she had gone away, and had become the wife of a white chief. She is now in your house. I come, therefore, white men, to demand justice. I first require that my wife be delivered up to me. She has acted like a dog and I shall live no more with her, but I shall punish her as she deserves. And in the next place, I expect, as you have been the cause of my losing her, that you will give me ample compensation for her loss."

Our interpreter immediately explained to the Indian that the girl's relatives were the cause of the trick that had been played on him. Had our friend been aware of her having been a married woman, he never would have thought of making her his wife.

He was willing to give him reasonable compensation for her loss, but she should not be delivered to him except he undertook not to injure her.

He refused to make any promises and still insisted on her restitution, but as we had reason to fear that her life would have been sacrificed, we refused to comply. The old chief next addressed him for some time. The result was that he agreed to accept a gun, one hundred rounds of ammunition, three blankets, two kettles, a spear, a dagger, ten fathoms of tobacco, with a quantity of smaller articles, and to leave his frail helpmate in quiet possession of her paleface spouse, promising never to think of her or do her any harm.

Exorbitant as these terms were, it was judged advisable to accede to them rather than disturb the good feeling that had hitherto subsided between us.

After we had delivered the above articles to him, we all smoked the "peace pipe". The fugitive, knowing that it was the ratification of peace, emerged from her place of concealment and boldly walked past her late lord. She caught his eye for a moment, but no sign of recognition appeared. Neither anger nor regret seemed to disturb the natural serenity of his cold and swarthy countenance.

12

WOLVES

THE PRAIRIE WOLVES ARE MUCH SMALLER than those which inhabit the woods. They generally traveled together in numbers. A solitary one is seldom met with. Two or three of us have often pursued from fifty to one hundred, driving them before us as quickly as our horses could charge.

Their skins are of no value. We do not therefore waste much powder and ball in shooting them. The Indians, who are obliged to pay dear for their ammunition, are equally careful not to throw it away on objects that bring no remunerating value. The natural consequence is that the wolves are allowed to multiply, and in some parts of the country are completely overrun with them.

The Indians catch numbers of them in traps, which they set in the vicinity of those places where their tame horses are sent to graze. The traps are merely excavations covered over with slight switches and hay and baited with meat, into which the wolves fall, and being unable to extricate themselves, they perish by famine or the knife of the Indian.

These destructive animals annually destroy numbers of horses, particularly during the winter season when the latter get entangled in the snow, in which situation they become an easy prey to their lightfooted pursuers, ten or fifteen of which will often fasten on one animal, and with their long fangs in a few minutes separate the head from the body.

If, however, the horses are not prevented from using their legs, they sometimes punish the enemy severly. As an instance of this, I saw one morning the bodies of two of our horses which had been killed the night before. Around were lying eight dead and maimed wolves, some with their brains scattered about, and others with their limbs and ribs broken by the hoofs of the furious animals in their vain attempts to escape from their sanguinary assailants.

While I was at Spokan I went occasionally to the horse prairie for the purpose of watching the maneuvers of the wolves in their combined attacks. The first announcement of their approach was a few shrill currish barks at intervals, like the outpost firing of skirmishing parties. These were answered by similar barking from an opposite direction, until the sounds gradually approximated and at length ceased on the junction of the different parties.

We prepared our guns and concealed ourselves behind a thick cover. In the meantime the horses, sensible of the approaching danger, began to paw the ground, snort, toss up their heads, look wildly about them, and exhibit all the symptoms of fear. One or two stallions took the lead, and appeared to wait with a degree of comparative composure for the appearance of the enemy.

The allies at length entered the field in a semicircular form, with their flanks extended for the evident purpose of surrounding their prey. They were between two and three hundred strong.

The horses, on observing their movement knew from experience its object and, dreading to encounter so numerous a force, instantly turned round and galloped off in a contrary direction. Their flight was the signal for the wolves to advance. Immediately uttering a simultaneous yell, they charged after the fugitives, still preserving their crescent form.

Two or three of the horses, which were

Owl Rock, photographed by B. A. Gifford in 1909. You can easily see two owls. The voyageurs paddled Cox by 17 times, but when he was back home in Dublin writing his book this pair of granite birds received no comment.

A pictograph near old Wishram Village called Witch's Head. The Klickitat bucks used it as a damper for coquettish squaws, who were solemnly told and believed that the eyes followed any unfaithful woman.

not in the best condition, were quickly overtaken by the advanced guard of the enemy. The former, finding themselves unable to keep up with he band, commenced kicking at their pursuers, several of which received some severe blows. These being reinforced by others, they would have shortly dispatched the horse had we not just in time emerged from our place of concealment, and discharged a volley at the enemy's center, by which a few were brought down.

The whole battalion instantly wheeled about and fled towards the hills in the utmost disorder. The horses, on hearing the fire, changed their course and galloped up to us. Our appearance saved several of them from the fangs of the foes. By their neighing they seemed to express their joy at our timely interference.

13

BEARS

BEARS ARE SCARCE ABOUT THE PLAINS but are found in considerable numbers in the vicinity of the woods and lakes.

Their flesh is excellent, particularly in the summer and autumnal months when roots and wild fruit are had in abundance. They are most dangerous animals to encounter, especially if they are slightly wounded, or that any of their cubs are in danger. In which case they will rush on a man though he were armed at all points, and woe to him if Bruin should enfold him in his dreadful grasp.

I have seen several of our hunters, as well as many Indians, who had been dreadfully lacerated in their encounters with bears. Some have been deprived of their ears, others had their noses nearly torn off, and a few have been completely blinded.

From the scarcity of food in the spring months they are then more savage than at any other season. During that period it is a highly dangerous experiment to approach them.

The following ancedote will prove this. Were not the fact confirmed by the concurrent testimony of ten or more, I would not have given it a place among my *memorabilia.*

In the spring of this year (1816) Mr. McMilan had dispatched ten Canadians down the *Flathead River* on a trading excursion. The third evening after quitting the fort, while they were quietly sitting round a blazing fire eating a hearty dinner of deer, a large half-famished bear cautiously approached the group from behind an adjacent tree. Before they were aware of his presence, he sprang across the fire, seized one of the men (who had a well-furnished bone in his hand) round the waist with the two forepaws, and ran about fifty yards with him on his hind-legs before he stopped.

His comrades were so thunderstruck at the unexpected appearance of such a visitor and his sudden retreat with *pouvre Louisson* that they for some time lost all presence of mind. In a state of confusion they were running to and fro, each expecting in his turn to be kidnapped in a similar manner. At length Baptiste Le Blanc, a half-breed hunter, seized his gun and was in the act of firing at the bear, but was stopped by some others who told him he would inevitably kill their friend in the position in which he was then placed.

During this parlay Bruin relaxed his grip of the captive whom he kept securely under him, and very leisurely began picking the bone which the latter had dropped. Once or twice Louisson attempted to escape, which only caused the bear to watch him more closely. On his making another attempt, he again seized Louisson round the waist and commenced giving him one of those infernal embraces which generally end in death. The poor fellow was now in great agony and vented the most frightful screams.

Observing Baptiste with his gun ready, anxiously watching a safe opportunity to fire, he cried out, *"Fire, pour l'amour du bon Dieu! A la tete! a la tete!"*

This was enough for Le Blanc, who instantly let fly and hit the bear over the right temple. He fell and at the same moment dropped Louisson, but he gave him an ugly scratch with his claws across the face, which for some time afterwards spoiled his beauty.

After the shot, Le Blanc darted to his comrade's assistance and with his *couteau de chasse* quickly finished the sufferings of the man stealer, and rescued his friend from impending death. With the exception of the above-mentioned scratch, he escaped uninjured.

They commenced the work of dissection with right goodwill, but on skinning the bear they found scarcely any meat on his bones. In fact, the animal had been famished and in a fit of hungry desperation made one of the boldest attempts at kidnaping ever heard of in the legends of ursine courage.

The skin of these animals are not at present held in the same estimation that they were formerly, particularly the brown or grizzly kind, few of which are now purchased. Good rich black ones and cubs still bring a fair price at the trading posts nearest to Canada and Hudson's Bay.

About twenty-five years ago the Company had a great number of bear-skins lying in their stores, for which there was no demand. One of the directors well known for the fertility of his expedients as an Indian trader, hit upon a plan for getting off the stock.

He selected a few of the finest and largest skins in the store, which he had made into a hammer-cloth [for hanging over the driver's seat of a coach] splendidly ornamented in silver with the royal arms. A deputation of the directors then awaited upon a late Royal Duke with the hammercloth and respectfully requested that he would be graciously pleased to accept it as a slight testimony of their respect. His Royal Highness returned a polite answer and condescendingly consented to receive the present.

A few days afterwards the King held a levee. His illustrious son proceeded to court in his state-coach with its splendid hammercloth. It attracted universal attention, and to every inquiry as to where the skins were obtained, the answer was, "From the North West Company". In three weeks afterwards there was not a black or even a brown bearskin in the Company's warehouse. The unlucky peer who could not sport a hammercloth of bear, was voted a bore by his more lucky brethren.

14

UP, UP THE COLUMBIA AND AWAY

THE SUMMER OF 1816 did not tend to diminish my growing aversion to the Indian country. Horse-racing, deer-hunting and grouse-shooting were pleasant pastimes enough, but the want of companionable society rendered every amusement "stale, flat, and unprofitable." Bad French and worse Indian began to usurp the place of English, and I found my conversation gradually becoming a barbarous compound of various dialects.

It was arranged I should pass the winter in my present post. I passed five weary winter months at Oakinagan without a friend to converse with. Tea and tobacco were my only luxuries, and my pipe was my pot-companion.

Towards the latter end of March, 1817, the other wintering parties joined us at Oakinagan, from whence we all proceeded to

Fort George, which we reached on the 3rd of April.

Wednesday, April 16, 1817. At one p. m. on this day we took our departure from Fort George under a salute of seven guns. Our party consisted of eighty-six souls.

On the morning of the 24th at three p. m. we arrived at the *Dalles* and immediately began the portage, but were only enabled to get half through it when we camped. We finished the portage at ten o'clock on the morning of the 25th.

On the 29th, shortly after sunset, we made our beds a little above the Wallah Wallah River. Tom Tappam, the chief, and several of his tribe, visited us and promised to trade some horses.

The 3rd of May we encamped at the foot of the Priest's Rapid. After breakfast on the morning of the 4th the party who were to cross the Rocky Mountains bade adieu to the loaded canoes and the gentlemen of the Columbia.

5th. About noon arrived at the portage of the Rocky Island Rapid. While Gingras and Landreville were getting one of the canoes up the rapid, the latter made a false stroke of his pole by which it missed bottom, and the canoe was upset in the middle of the waves. Gingras held fast to the bars until it was drawn into an eddy, when he found bottom and got ashore. In the meantime eight men leaped into the other canoe and instantly pushed off to the assistance of Landreville, who was for a couple of minutes invisible. At length appearing above the surface, they seized him by the hair and drew him on board nearly lifeless. All of our baggage was subsequently picked up. We remained here the remainder of the day to dry it, and repair the canoe.

On the 8th we reached Oakinagan Fort, where we passed the night.

14th. On arriving at the Grand Rapid we were forced to carry the canoes as well as the baggage to the upper end. This occupied the greater portion of the day. We did not finish before three p. m. At four we arrived at the Great Kettle Falls, the portage of which we completed at sunset.

On the 15th made pretty good way until one p. m. when we arrived at a particular part of the river, called the First Dalles or narrows, above the Kettle Falls, where the channel is confined between a range of high and dangerous rocks nearly a mile in extent, the whole of which distance the men were obliged to carry the canoes and baggage. The

Indians assert that no rattlesnake are to be found on either bank of the Columbia above this river, and all our men who had been previously in the employment of the company, hunting in that part, fully corroborated this statement.

About seven o'clock on the morning of the 16th we passed the mouth of the Flathead River, which falls into the Columbia over a foaming cascade. During the day we passed a number of small rivers, which, owing to the melting of the snow, had swollen into torrents. The force of the current rushing out from these rivers, repeatedly drove the canoes back with great violence, and it required all the skill and strength of our men to pass them.

About an hour before we encamped we observed a large black bear in the act of swimming across the river, which Mr. McGillivray wounded. The enraged animal instantly changed its course and came in contact with our canoe, into which it attempted to get by seizing the gunwale with its forepaws. This nearly upset us, but the foreman aimed a well-directed blow at his head with his pole, which completely stunned it, and we succeeded in hauling it on board. It was in rather good condition, and proved a welcome and unexpected treat.

17th. Entered the first lake formed by the Columbia, between eleven and twelve leagues long and about one and a half in breadth. In the distance we first caught a view of the most western chain of the Rocky Mountains covered with snow. A headwind during the greater part of the day considerably retarded our progress.

On the morning of the 18th we entered that part of the river called the Straits, which separates the Upper from the Lower Lake. It is only a few miles in length and quickly brought us to the upper lake which is not so long as the first. Encamped at sunset at the upper end of the lake on a fine sandy beach. During the day we struck on two sandbanks, and were slightly injured by a sunken tree.

19th. About two miles above our encampment of last night the Columbia becomes very narrow, with steep and thickly wooded banks covered with immense quantities of fallen trees. The current is very strong and, owing to the great height of the water, the men at intervals had scarcely any beach on which to walk in dragging up the canoes.

At nine o'clock on the morning of the 20th we reached the Second Dalles. There are high and slippery rocks on each side, which makes it a work of great danger and difficulty to pass them. The baggage was all carried by the men. The canoes were towed up with strong lines, after being in great danger of filling from the frightful whirlpools close along the shore. From dawn of day until noon on the 21st we did not make three miles, owing to the impetuosity of the current, in the shelving banks, and the extreme weakness of our men. We were detained at one place upwards of four hours to repair our shattered canoes.

22nd. About two p. m. arrived at the Upper Dalles. Got about halfway through this channel, and stopped for the night in a small nook formed by the rocks, on which we lay scattered and exposed to severe rain during the night.

We rose wet and unrefreshed on the morning of the 23rd, and in five hours passed the Dalles, the upper part of which of a chain of whirlpools, which compelled us to carry both canoes and baggage some distance over the rocks. Some of the men narrowly escaped with their lives. Those who carried our canoe, from mere exhaustion fell several times, by which the canoe was damaged. We were detained until three p. m. to get it repaired.

encamped at dusk on a sandy beach. The rain continued during the evening, and the night to pour down in torrents.

Our progress on the 24th was equally slow. The various tributary streams swollen into torrents, as they rushed with headlong force into the Columbia, repeatedly drove us back with irresistible strength, and at times we were in danger of filling. On two occasions, where the opposite shore of the Columbia consisted of perpendicular rocks, we were obliged, after various fruitless attempts to pass the minor streams, to unload and carry the canoes some distance along their banks until we reached a smooth space of current, when we crossed and by that means surmounted the difficulties. It rained on us all the afternoon.

On the 26th we only made three miles, in the course of which our canoe filled in a dangerous rapid and we were near perishing. We succeeded, however, in gaining a low stony island on which there was no wood to light a fire. Our pemmican was completely damaged by the accident. As a climax to our misery, it rained incessantly the whole day.

The river here opened out to a considerable breadth, and in some places was very shallow. The Rocky Mountain portage at which we were to leave our canoes, appeared in sight, not more than three miles distant. As we threw our jaded bodies on our stony couch this evening we most truly experienced that "weariness can snore upon the flint".

On the 27th at nine a. m. arrived at the entrance of Canoe River, where the portage commences. With indescribable pleasure we bade final adieu to our crazy battered canoe.

Messrs McDougall and Bethune had reached it the day before, and had almost despaired of seeing us. Finding so many of our men invalids, those gentlemen deemed it imprudent to bring them across the mountains. Six Canadians and Homes the English tailor were therefore sent back in the best canoe to Spokan House. It was hoped they would arrive in three days at the Kettle Falls, from whence they could easily reach Spokan. As our stock of provisions was very scanty, we could only spare them enough for the above period.

Our baggage and provisions were divided between the nine remaining men, who were obliged to carry about ninety pounds weight each, besides their own kits which in such cases are never taken into consideration.

September 19th. Drove off for Montreal, in which city we arrived at half past nine

p. m. at Clamp's Coffee House in Capital Street, after a journey of five months and three days from the Pacific Ocean.